The Arnott was produced by Daphne Arnott's company, which also ventured into the production of road and sports racing cars. The Arnott enjoyed success, driven by the likes of Ivor Bueb and Dennis Taylor.

The 500cc Racing Car

Colin C. Rawlinson

A Shire book

Contents

ACKNOWLEDGEMENTS
Illustrations are acknowledged as follows: Bugatti Owners' Club, pages 11 (bottom), 22; Ferret Fotographics, pages 3, 4 (top), 6 (top and centre), 8 (bottom), 9 (both), 11 (top and centre), 13 (both), 14 (top and centre), 15 (both), 17 (centre), 19, 20 (bottom), 25, 26 (top and bottom), 28 (centre), 29 (bottom); the 500 Owners' Association, pages 1, 4 (bottom), 5 (top), 7 (bottom), 8 (top), 14 (bottom), 17 (top and bottom), 21, 24, 27 (top), 29 (top), 31 (bottom), 32; Derek Hibbert, front cover; Roy Hunt, page 30 (top); Kane Products Ltd, page 18; Hugh Langrishe, page 28 (bottom); G. Maclean, page 10; National Motor Museum, pages 6 (bottom), 7 (top), 16, 20 (top), 26 (centre), 30 (bottom), 31 (top); Thomas Nelson & Sons Ltd/Raymond Groves, pages 12, 28 (top); Dick Tudhope, page 27 (bottom). The photograph on page 5 (bottom) is from the author's collection.

Cover: *Mike Lawrence attacks Shelsley Walsh hill-climb in his Cooper in this twenty-first-century scene.*

British Library Cataloguing in Publication Data: Rawlinson, Colin C. The 500cc racing car. – (Shire album; 417) 1. Automobiles, Racing – History 2. Automobile racing – History I. Title 629.2'28. ISBN 0 7478 0555 5.

Editorial Consultant: Michael E. Ware, former Director of the National Motor Museum, Beaulieu.

Published in 2003 by Shire Publications Ltd, Cromwell House, Church Street, Princes Risborough, Buckinghamshire HP27 9AA, UK. (Website: www.shirebooks.co.uk)

Printed in Great Britain by CIT Printing Services Ltd, Press Buildings, Merlins Bridge, Haverfordwest, Pembrokeshire SA61 1XF.

Early days of the movement

The concept of a formula for full-scale racing and competition for 500cc cars originated just after the Second World War. The idea was simple enough: to formulate a set of rules that would provide cheap motor-racing in the early post-war years and provide an opportunity to compete for many motor-racing enthusiasts who would otherwise never have stood a chance of participating in organised competitive motor sport.

This initiative was taken up and put into practical effect by a group based in Bristol, but it was not very long before the idea took hold across Britain and resulted in the formation of the '500 Club', later renamed the 'Half-Litre Club'. The rules were few and straightforward, but the fundamental one was that the engine capacity for the new single-seat racers was limited to 500cc unsupercharged. As there were no car engines suitable for the purpose, the new formula effectively prescribed the use of a motorcycle engine as the power unit. This was a different matter; these were in plentiful supply as most of the motorcycle manufacturers (of which there were a fair few at that time) had a 500cc model whose engines could be used. However, although numerous types were tried initially, the requirements of performance and reliability pointed virtually all the constructors in one direction – the speedway JAP. J. A. Prestwich of Tottenham was in fact a manufacturer not of motorcycles but of proprietary engines designed for certain specific purposes, one of which was for speedway – or dirt-track – motorcycle racing. Consequently there was available a unit that was designed from the outset as a performance engine and required little in the way of tuning to produce results. As the construction and series production of 500cc racing cars progressed, the JAP gave way to the racing Norton; but we are getting

Jeff Sparrowe at the Prescott (Gloucestershire) hill-climb in 1948 in his own creation, the SMS. Another Austin Seven-based special, the SMS was fitted with a Rudge Ulster engine, rear-mounted as had become the norm.

An early one-off 500 – 'Buzzie', built by Jim Bosisto of Bristol. 'Buzzie' is seen here rounding Pardon hairpin at Prescott in 1948, driven by John Ebdon. Note the sylvan surroundings of Prescott at that time – very different from the Armco and wide-open spaces of today. Ebdon subsequently rebuilt this car into the Halldon-JAP.

a little ahead of the story. The underlying principle of the small capacity was to make motor-racing available to many enthusiasts at a modest cost and to open the door of opportunity to many who would otherwise, for financial reasons, never have aspired to participation in any form of motor sport.

The early pioneers of the movement – those who were the first actually to construct a vehicle that could perform in competition – have now passed into legend. Devotees of 500cc racing will be familiar with the names Colin Strang, Clive Lones, John Cooper and Eric Brandon. Their successes resulted in enthusiasm for the new formula spreading with great rapidity, and it was not long before numerous other cars joined in the competition. One of the important principles of the movement, certainly in its initial stages, was to provide encouragement for the home constructor. Numerous 'specials' began to appear and these early one-off creations gave enormous impetus to the movement. One of the first moves towards serious production came when several enthusiasts, anxious to join the fray, asked John Cooper if he would build a car for them. The first Cooper, the product of John and his father Charles, made its first appearance at the Prescott hill-climb in 1946 and quickly became a great success; a similar car was produced for John's friend Eric Brandon. From these beginnings the Cooper garage at Surbiton, Surrey, went into

An early picture of Wally Cuff at Naish Hill near Bristol in 1952. His Iota-based 'Hell's Hammers' is a good example of an early home-built special. Naish was unusual in that it was a grass-surfaced hill and not a metalled road.

The Cooper works c.1953 with part-completed cars on the production line. The twin-tubed chassis frames indicate Mark 6 or Mark 7.

production on a small scale and before long its output had rapidly expanded. One of the first customers wanting a Cooper 500 was a young man called Stirling Moss, who took delivery of one of the early productions almost as soon as he had passed his driving test. The success of Stirling Moss, who went on to become one of the greatest racing drivers of all time, is well known. Moss was one of several 500 drivers who progressed to international stardom.

By the following year, 1947, the Half-Litre Club had grown enormously in membership and was even producing its own magazine. The movement spread to the Continent and a

Covers of 'Iota', the Half-Litre Club magazine. The earlier one (August 1947) shows Colin Strang (note the absence of a crash helmet). At that time the magazine was the official organ of the 500 Club, which, as can be seen on the later magazine cover, was renamed the Half-Litre Club. The title 'Iota' is derived from the Greek version of the letter 'I', denoting the category in which 500cc cars were placed by the international racing organisation, the FIA.

Left: *Continental venture – Don Parker's Parker-CFS is seen in the pits at Rheims on 2nd July 1950. Formula Three was an area in which British drivers excelled and could rarely be matched by Continental opposition. Parker's car was a home-built machine, beautifully engineered by Parker himself and his associate Charlie Smith, a well-known motorcycle exponent. Parker enjoyed several successful Continental outings and on this occasion finished fourth in the Robert Benoist Challenge race.*

Right: *Stuart Lewis-Evans met with spectacular success in 500cc racing and went on to become a Grand Prix driver for Vanwall. Sadly he was killed at the Moroccan Grand Prix in 1958, the race at which Mike Hawthorn clinched the World Drivers' Championship. The Lewis-Evanses, father and son, both participated in Formula Three although 'Pop' Lewis-Evans is better known as 'team manager'. Stuart drifts his Cooper-Norton round a Goodwood bend in this shot.*

Left: *Jim Russell, a garage owner from Downham Market in Norfolk, first witnessed 500 car racing at Snetterton, the track nearest his home, and was so impressed that he was anxious to join the action right away. He started with a Cooper JAP in 1952 but subsequently had his car fitted with a 'double-knocker' Norton engine prepared by the tuning wizard Steve Lancefield. He immediately became a force to be reckoned with and chalked up numerous successes, enjoying a keen rivalry with drivers such as Don Parker and Ivor Bueb. He is ready for action at Boreham in this 1952 picture.*

Aintree, September 1953. Les Leston is receiving 'technical adjustments' in his Cooper prior to the start of the 500 race at The Daily Telegraph Trophy meeting.

number of cars were constructed to the 500cc formula – although it has to be said that the enthusiasm of continental constructors never matched that found in Britain. In 1950, the 500cc formula was accorded the international status of Formula Three in recognition of its rapid rise in popularity (Formulas One and Two were already in existence and covered different categories of racing car). The movement spread worldwide – a number of Australian 500s were built and raced – but most of the organisation and activity was centred in Britain and the rest of Europe. It was encouraging to see British drivers dominating international Formula Three racing, a pleasant change from the situation that prevailed higher up the motor-racing scale.

One of the most successful of all the 500 drivers was Don Parker, who started his racing career with his own special, with which he achieved third place in the Prix de Monte Carlo, held in 1950, a curtain-raiser to the Monaco Grand Prix itself. Parker went on to win numerous championships with a Kieft but there were a number of other very successful 500 drivers, including Les Leston, Stuart Lewis-Evans, Jim Russell and Ivor Bueb. The cars were also used for sprints and hill-climbs and gained many successes in these fields.

This picture of Don Parker shows him winning his last Formula Three race, at Crystal Palace in August 1959. By this time even Parker had changed his allegiance from Kieft to Cooper. He was the most prolific winner of 500cc races ever, chalking up 126 wins in all. He retired at the end of the 1959 season.

Another leading Formula Three driver – Jim Russell in his Cooper Norton. A prolific winner of Formula Three races and a national champion, Russell was one of the several users of 'double-knocker' Norton engines tuned by Steve Lancefield. This Goodwood shot was taken in September 1957.

When 500cc racing reached the peak of its popularity, races were held all over Britain at numerous circuits and spectators witnessed much extremely close and exciting racing from the diminutive cars, with over sixty entries being not unknown at some meetings. In other countries, as already indicated, British drivers picked up virtually all the major awards and the Cooper marque enjoyed almost runaway success.

By the late 1950s the popularity of Formula Three was beginning to wane for a number of reasons. Public enthusiasm and interest had been catered for by the excellent racing and

One factor that was favourable to motor-racing after the Second World War was the abundance of redundant wartime airfields, a number of which incorporated a perimeter track that could be used for racing. One such airfield was at Ibsley near Ringwood in Hampshire. The first race meeting was held at Ibsley in August 1951 and the second, depicted here, in April 1952. The cars leaving the line in the final of the 500cc are Andre Loens (Kieft-Norton 50), John Habin (Staride-Norton 47) and Charles Headland (Kieft-Norton 61). Coming into shot is John Coombs (Cooper-Norton). The winner of the race was Headland. Note the splayed rear wheels on all three of the cars in the front row. This reflects the suspension system of elastic bands and swing-axles used to obtain independent rear suspension.

Formula Three was well known as a means by which a number of drivers who started in this form of racing went on to become international stars (Stirling Moss, Stuart Lewis-Evans and Peter Collins among others), but occasionally an established top-line driver stepped down the ranks to try his hand at Formula Three. One such was Bob Gerard (Cooper 45), seen here at Goodwood in the Daily Graphic Trophy Meeting on 27th September 1952. He was not placed on this occasion. Also in the picture are Paul Emery (Emeryson 66), Andre Loens (Kieft 57), John Brise (Cooper), Reg Owen (Hill), Austen Nurse (Cooper) and Bill Aston (Arnott).

entertainment provided by the 500cc racing car, but nothing stands still in technology and a new form of racing had arrived – Formula Junior. This took over from 500cc racing as the mainstay of single-seater competition at that level and the 500cc cars were sidelined into minor events. Formula Junior provided for what were in effect scaled-down Grand Prix cars, using suitably modified production engines up to 1100cc, and was able to incorporate much in the way of up-to-date technological development.

One of the stalwarts of 500cc racing almost from the start was Don Truman, who drove a succession of different cars, starting with a Marwyn, which was rebuilt into the Bardon special. This subsequently became the Bardon-Turner. Truman then became a Cooper driver with the purchase of a Mark 6. He is seen here leading David Boshier-Jones (Kieft) at Castle Combe in October 1953.

Among the first cars constructed for the new 500cc formula in 1946 was Clive Lones's home-built 'Tiger Kitten'. One of the few to have a front-mounted engine, it was remarkably successful in the early years and Lones was the first driver to record under 50 seconds for an ascent of the half-mile-long Prescott hill.

The cars described

As we have seen, the 500cc racing movement could not have gained the popularity it did without the original band of enthusiasts who set about producing a car to the new 500cc formula in the years immediately following the Second World War. A great deal of ingenuity went into the early creations, many of whose builders turned to the Austin Seven and Fiat Topolino chassis to provide suitable bases for the construction of a 500cc car powered by a motorcycle engine. These had the advantages of being of a suitable size and weight and were readily obtainable at a modest cost. One of the very first builders on the scene was Clive Lones, who had raced a Morgan before the war. His first 500cc racer, 'Tiger Kitten', was based on an Austin Seven chassis obtained from a scrap dealer. A JAP engine was installed initially but, in the manner of many a special, constant modifications were employed to obtain progressively greater performance and it was a fitting tribute to Lones's perseverance that his car was the first 500 to record a time of under 50 seconds at the Prescott hill-climb.

Another early and successful 500 was built by Harrow motor engineer Colin Strang. He was a man with truly original ideas. Strang hit on the notion of building his 500cc special by locating the engine behind the driver using a motorcycle clutch, gearbox and chain transmission. The principle of a rear-engined car was not new – it had been successfully used by the great German Auto Union Grand Prix team before the Second World War – but it was Strang who achieved immediate post-war success with this layout in such a diminutive car. A Vincent-HRD engine was used and the Strang became a familiar sight at most of the events organised for 500cc racers in the early years.

Contemporary with Lones and Strang, a number of other enthusiasts built cars in the

A true pioneer: one of the first to produce a 500cc car in 1946 was Colin Strang, seen here at the Prescott hill-climb in 1946. Strang used a Fiat Topolino chassis as the basis of his special. A rear-mounted Vincent-HRD motorcycle engine powered the car driving via motorcycle gearbox and transmission. The Strang proved to be a very successful car and set numerous records in the early days.

early stages of the 500cc movement. They may not have achieved great success but nevertheless an enormous debt is owed to them for being part of the nucleus of a formula that spread like wildfire. Frank Bacon, Wing Commander Frank Aikens, Charlie Smith, Gerald Spink and Bill Grose may not have been household names but they all played their part in contributing to the subsequent success of 500cc racing.

It is generally believed that John and Charles Cooper, impressed by the practicality and proved performance of the Strang, consequently adopted the rear-engined layout, which

Right: *Displaying all the characteristics of a home-built special, John Ebdon's Halldon-JAP is seen at speed in the 500 race at Lulsgate, Bristol, in April 1950. Ebdon produced this car from the original 'Buzzie' (see page 4) and with it put up a number of most respectable performances. He finished fifth on this occasion from an entry of fourteen cars.*

Left: *A driver who typified the original spirit and enthusiasm of the 500 movement was Frank Bacon, builder of the FHB, in which he awaits the signal for the 'off' at Prescott in 1948. This is an excellent example of an amateur-built 'special' from the early years. Bacon's first effort was a front-engined machine based on an Austin Seven chassis. This was rebuilt into a rear-engined car, which is the one we see here.*

became the norm for virtually all cars constructed to the formula. Indeed, the subsequent success of the Cooper Car Company, which went on to become a pioneer of the modern rear-engined Grand Prix racing-car design and to win world Grand Prix championships, set the path followed by racing-car designers, and it can be argued that Colin Strang should rightfully be regarded as the father of the modern Grand Prix car.

Cooper had already commenced its production run and had built a significant number of cars by October 1948, when a major event took place in the history of 500cc racing – the first full-scale race over a substantial distance. The RAC National race took place over a distance of 50 miles (80 km) and no fewer than thirty-four entries were accepted. The race was run as a curtain-raiser to the British Grand Prix at Silverstone. The entry and grid typified the wide variety of cars that had by then been constructed, and the race provided a severe test for all the machines. Many individual dramas took place before the grid assembled, which, in its final form, comprised twenty-six cars: nine Coopers and seventeen 'specials' constructed by amateur enthusiasts. This occasion provided a splendid opportunity for the new formula to be seen in numbers by the public at large and the race proved extremely popular. The rate of failure was, however, rather alarming; of the twenty-six cars that started, eight were still running at the finish, and of those eight five were Coopers. The winner was a colourful character by the name of Spike Rhiando, who covered the 50 miles (80 km) at an average speed of 60.68 mph (97.63 km/h). He led home John Cooper, Sir Francis Samuelson and Eric Brandon, these first four all being Cooper-JAP mounted. R. L. Coward, in the Cowlan – which was later to become the basis of the prototype JBS – was the first of the amateur-built cars to finish, in fifth place.

Winner of the famous 500 race at Silverstone in October 1948, Spike Rhiando is shown in this contemporary illustration at the wheel of his Cooper.

By 1953 most drivers favoured factory-built machines and the one-off home-built specials were no longer competitive. This was as inevitable as it was sad but, none the less, some competitors continued to have their fun with machines of their own creation. Here is Peter Trevellick at Silverstone in May 1953.

The technical details of the various specials were extremely varied although they nearly all followed the rear-engined layout, using a motorcycle engine, clutch and gearbox with chain transmission. The rate of failure is not surprising when it is borne in mind that until that time most of the cars had been used for events over much shorter distances and were untried in any form of endurance racing. A variety of engine manufacturers was represented, namely JAP, Rudge, Norton, Scott, BSA, Vincent-HRD and Triumph.

Following the success of the 1948 Silverstone race, the 500cc formula captured the public's imagination and experienced a tremendous upsurge in popularity, with Cooper increasing its production, many more individually built specials appearing, and a number of other manufacturers going into series production. Practically all cars produced for 500cc racing thereafter followed the now established basic layout of the engine behind the driver and transmission by means of a primary chain to the gearbox and clutch and a secondary or drive chain to the rear axle. This set-up gave rise to many variations in individual design involving chassis, suspension, rear-axle layout and numerous other features aimed at increasing performance and road-holding. There were a few exceptions to the conventional layout, notably the Emeryson: a small factory produced a total of eight of these beautifully designed front-engined machines, in which the principle of motorcycle transmission was employed, with the chain driving the front wheels. Four of the eight cars were still in existence in 2003.

Ken Smith first started driving a 500 in 1948 with his own special. After this came to grief at Brands Hatch in 1950, Smith commissioned a chassis from Buckler of Reading, an example of a one-off car being professionally built. With it he won the non-production race at Silverstone in August 1951. He is seen here at Snetterton in July 1953. This car was acquired by the author in 1985.

Welshman David Boshier-Jones became a very successful Formula Three driver. Although he enjoyed a fair amount of success on the circuits it was in hill-climbs that he made his mark, being RAC champion with a hat trick of successive wins in 1958–60. However, the latter success was achieved with a 1100cc-engined Cooper. He is seen here in a Kieft at Castle Combe in October 1953.

Right: *Reg Bicknell of Southampton was an innovative designer who decided to apply aerodynamics to the bodywork of his car. He made his first racing appearance at Silverstone in August 1951. Appropriately, this picture shows the Revis at Ibsley, Bicknell's 'local' track.*

The Formula Three field roars away from the start at Snetterton in August 1954. On the front row are Don Parker (Kieft), Jim Russell (Cooper), Ivor Bueb (Cooper) and Eric Brandon (Cooper). Behind them can be seen Tommy Bridger (Kieft), Reg Bicknell (Revis) and Paul Swaelens (Cooper). Bob Gerard is in the Cooper with the white nose, to the right of Bueb. The race was won by Jim Russell at an average speed of over 80 mph (129 km/h).

Ibsley, April 1952 – Andre Loens (a Belgian living in Britain) waits on the grid in his Kieft with Don Parker (Kieft 59) and Tom Clarke (CBP 55) in the background. Loens came third in his heat but retired in the final, having started on the front row. Note the open-necked shirt and rolled-up sleeves!

A number of manufacturers went into production to a greater or lesser extent (see Appendix). With the Cooper factory at Surbiton producing an ever-increasing number of cars, the gradual eclipse of the home-built special became inevitable. The most successful of the manufacturers to rival Cooper were undoubtedly JBS and Kieft. The JBS was distinctive for its innovative and cleverly designed rear suspension. Apart from this, its build followed well-tried principles and it proved to be a very fast car that was more than capable of challenging the Coopers. The JBS project was masterminded by Wembley speedway rider Alf Bottoms but was unfortunately cut short by Alf's death in an accident

Andre Loens's Kieft at Castle Combe in October 1952. The driver sat very far forward in the Kieft when compared with the Coopers.

Don Parker scored numerous Formula Three wins in his career, mainly with the Parker-Kieft. Although originating from the Kieft factory, the car was extensively modified to his own specifications. This picture shows Parker at the start of the Earl of March Trophy at Goodwood in April 1956, his wife Dora standing beside the car. To the right of the picture is the Cooper of Ken Tyrrell.

in Luxembourg. The Kieft, however, was a make that continued to put up a very successful resistance to Cooper domination. It was produced by Cyril Kieft and his team, originally at Bridgend in South Wales and latterly at Wolverhampton. Cyril Kieft entered the 500 scene after purchasing a Marwyn, the first of the production 500s. After persevering with his own designs, including successful international speed record attempts at the Montlhery track in France, he came to an arrangement with Stirling Moss to acquire the prototype and production rights to a revolutionary new design. The most distinctive feature of this Kieft was its suspension, which made extensive use of rubber combined with wishbones at the front and rear swing-axles. Stirling Moss remained as driver and with him at the wheel the new Kieft was every bit a match for the Coopers, inflicting numerous defeats on the latter. Another extremely successful 500cc driver, Don Parker, acquired two Kiefts, which he modified extensively. With these machines Parker also enjoyed enormous success. He had another advantage too – he weighed only 8 stone (51 kg)!

Meanwhile, John and Charles Cooper had not been sitting idly back; they produced a succession of new models, each one incorporating improvements on its predecessor and drawing on lessons learnt on the track. It is interesting to compare the original prototype, which appeared at Prescott in 1946, and the last new Cooper design, the Mark 9, later versions being virtually unchanged from this, although even this last design underwent a little evolutionary process of its own. The prototype Cooper basically consisted of two front sections of a Fiat Topolino chassis placed back to back and welded together, the cunning objective being to obtain the benefit of independent springing at both the front

Eric Brandon pressing on in one of the two prototype Coopers – here at Prescott in 1948.

Right: *A Cooper exposed – a large number of typical features of a 500cc racing car can be seen in this picture. These include side-mounted fuel tank, oil tank, final drive sprocket, primary and drive chains, hydraulic brake pipes and Norton overhead valve engine.*

David Boshier-Jones in his Cooper Norton at Goodwood in September 1957. He finished fourth in the 24 mile (39 km) race.

and the rear. The wheels were also taken from a Fiat Topolino, as was the steering. The faithful Speedway JAP engine was coupled to a Triumph motorcycle clutch and gearbox. The regulations provided for bodywork being 'optional but desirable' and a smart aluminium body gave the car a finished appearance. This prototype could rightly be regarded as a 'home-built special' as the car was originally built as a one-off with no intention of a subsequent production run. In its debut at the Prescott hill-climb the car suffered major problems by having engine mountings too weak to withstand the stresses

Contemporary 'cigarette cards' by Kane Products (a confectionery company), which issued a series of cards depicting the motor-racing scene, including several of 500s. (Clockwise from top right) Stirling Moss (Kieft); the Coopers of Ecurie Richmond, usually driven by Alan Brown and Eric Brandon; a typical 500 racing scene; Stirling Moss (Cooper); Don Parker (Kieft); Les Leston (Leston Special).

produced by the JAP engine and required improvised modifications to overcome this problem before the car was able to make a successful run.

By contrast, the last model produced by Cooper, in the mid 1950s, was a very different machine. It incorporated what were at that time the latest design and techniques and was almost unrecognisable from its humble origins. Just about every aspect of the car's construction was different from its prototype, other than its basic layout. The Mark 9 was built with a lightweight tubular chassis, light alloy wheels, up-to-the-minute steering, suspension and braking, and was powered by the famous Manx Norton twin overhead camshaft racing engine. It is notable that the Cooper design evolved so dramatically in the relatively short period of under ten years.

Revival

We have seen that 500cc racing experienced a very rapid rise to popularity in a remarkably short time to reach its peak in the years of 1951–5. Formula Three circuit racing did continue for some years after this but inevitably in due course gave way to the attractions of other formulae such as the new Formula Junior, which had the advantage of being able to draw upon technical improvements and the developments in motor-racing in general. While taking nothing away from the achievement of the Coopers, the greater numerical presence of cars of this marque and their increasing reliability because of good series-production techniques predictably resulted in Cooper domination of the formula. The ultimate development of the Cooper coupled with the outstanding performance and reliability of the Manx Norton power unit resulted in a situation familiar to all motor-racing followers; the best cars attracted the professional drivers, which meant that the enthusiastic amateur driving anything other than one of these machines was almost certainly destined to be an 'also ran'. Many of the top drivers began to drift away from Formula Three racing to greater things. 500cc racing had not only proved itself successful and popular in its own right; it also provided an entry into motor-racing for a number of drivers who commenced their careers in the formula and went on to become international stars. Such successes include Stirling Moss, Peter Collins, Stuart Lewis-Evans, Les Leston, Ivor Bueb, Jim Russell and Trevor Taylor. Other notable motor-sport luminaries such as Bernie Ecclestone and Ken Tyrrell also drove regularly in Formula Three racing.

Thus by 1960 500cc racing was all but dead and circuit racing virtually discontinued. The cars themselves suffered a variety of fates, from being scrapped to being shunted into

Goodwood circuit in Sussex staged many epic 500cc races and was a very popular venue for Formula Three. At the start of the first heat of the 500 International Trophy held at Whitsun 1950, Peter Collins (Cooper Norton 5) gets away from the front row alongside Alan Brown, also Cooper mounted, and Bob Cowell, driving Paul Emery's front-engined Emeryson. Collins went on to finish second in the final.

Brands Hatch circuit in Kent was ideally suited to 500cc racing and became one of the most popular venues in Britain. Initially the races were run in an anticlockwise direction but after a few meetings this was discontinued and the traditional clockwise direction for motor-racing circuits was adopted. A typical 500 field gets under way in April 1958. Prominent are Menzies (Petty Norton 11) and Koring (Smith 14).

oblivion and stored in barns and outhouses. A notable few were kept running in speed events such as sprints and hill-climbs; however, these rarely competed in their own separate class but rather, with little chance of success, against larger capacity cars. To those few owners who kept their cars running in the 1960s a debt is owed. In 1969, a group of enthusiasts joined forces and decided that an association ought to be formed for owners of 500cc racing cars. They felt that the unique place occupied by 500s in motor-racing

The 500cc car movement has much to thank this man for. He is Peter Kendall, who was instrumental in the formation of the 500 Owners' Association in 1969. The movement has since gathered momentum and has members in many countries. This picture shows Kendall in his Cooper-Norton at the Prescott hill-climb in 1967.

George Saunders in the first production Cooper 500, a Mark 2, at Prescott in 1948. Note the first use of the typical Cooper cast wheels and reversed inclination of rear shock absorbers.

history should not be allowed to pass into obscurity and so they set about contacting all known owners and spread the word to all those interested in keeping the cars active. The group, headed by Peter Kendall, brought about the formation of the 500 Owners' Association, with membership open to anyone interested in the cars whether an owner or not. Since its inception, the Association has been responsible for promoting active participation in motor-racing events and, although competition has been confined largely to sprints and hill-climbs, the resurgence of interest and activity in the old cars is beyond any doubt. Many models have been traced and lovingly restored; a good number have now passed their fiftieth anniversary. In addition, the Association stages exhibitions and demonstrations and offers a technical and spares service to its members. There is always great excitement when a long-forgotten car is 'discovered' and brought back to life. Some of the very early cars are now in the hands of devoted enthusiasts – the Strang, the Squanderbug, the 'Buzzie', the Bardon-Turner and several of the earliest Coopers being notable examples. One of the Don Parker Kiefts emerged from immaculate restoration by a member of the Association.

The revival movement has continued to provide close and exciting racing. Stars who have enjoyed 'modern' successes with these historic cars are Julian Majzub (Cooper), John Turner (Cooper Mark 9), Rodney Delves (Kieft) and Bill Needham (Cooper Mark 4) on the circuits, and Colin Myles, Ron Warr, father and son duo Tim and Ewan Cameron, Jon Brough, Barry Brant, Jonathan Docherty (all in Coopers) and Richard Neale (in the one-off Smith) on the hills. In 2001 Julian Majzub achieved his fourth successive 500cc race at the annual Goodwood Circuit Revival meeting, although Reg Hargrave in a Kieft was the winner in 2002.

There is no sign of the popularity of these cars waning and the revival of Goodwood circuit and its featuring of a 500cc Formula Three race with over thirty entries from the United Kingdom, Sweden and elsewhere has heralded a resurgence of interest on the tracks. One interesting result of the renewed activity is the number of elderly spectators who take delight in seeing the old cars in modern competition, having seen them racing 'first time round' in their youth. Many is the time when the driver of a car is approached by a member of the public with words such as 'I remember seeing your car driven by so and so in 1954 at such and such a track'.

500cc car racing outside the United Kingdom

As 500cc racing originated in Britain, much of its history and development took place there and the events described in this book are largely centred around the British racing scene. However, the principle of cheap motor-racing in the years immediately following the Second World War was not entirely a British preserve as, not surprisingly, the concept appealed to many enthusiasts on the Continent and, indeed, worldwide.

As in Britain, numerous one-off amateur-built cars made their appearance on the Continent and races were organised in several European countries. In both France and Germany there were attempts at series-production 500s but one of the most enthusiastic countries to embrace the formula was Sweden. 500cc races were held all over the Continent; some were largely local affairs but others were billed as international events and attracted many a British 'invader'.

While the initial enthusiasm for constructing 500s almost matched that in Britain, interest outside the United Kingdom was not maintained as it was within Britain. Overall, the 500cc movement spanned roughly a fifteen-year period (1946–60). Those Continental races that attracted British entrants were usually won by one of their number, such was the experience and competitiveness of the British driver. While locally built machines enjoyed

In July 1952 a hill-climb meeting was held at Prescott exclusively for 500cc cars. The meeting attracted a number of foreign drivers, among whom was Helmut Deutz from Germany, driving his DKW-engined Scampolo, surely the only appearance of a German 500 at a British hill-climb event.

a fair measure of success in club events abroad, the winner often turned out to be a local man who had simply imported a British Cooper!

Nevertheless, there were some useful Continental cars, for example the French DB with its Panhard engine. Germany produced cars such as the Scampolo and Monopoletta. The Swedish Effyh also enjoyed some success and one of these JAP-engined cars, in the hands of Ake Jonsson, finished second in the Challenge Robert Benoist at Rheims in July 1950.

There were occasional incursions, too, of foreign cars into British events. Unfortunately they could do little to make any impression on the rapidly developing British cars. A regular Continental competitor in British events was Dutchman Lex Beels, driving his own Beels 500. In 1952, the Prescott hill-climb held an international meeting confined exclusively to 500cc cars. Some interesting continental 500s made an appearance at the event, notably Beels cars driven by Lex Beels and Pim Richardson, and Helmut Deutz's German-built Scampolo, which used a DKW engine, driven by Deutz himself. The foreign visitors put up a good showing considering they were not familiar with the short style British hill-climb course, and although they did not feature as class winners, their appearance was none the less very welcome.

Numerous home-built 500s with a sprinkling of imported Coopers also provided some good motor sport in parts of the Commonwealth such as Australia, New Zealand and South Africa. Ron Tauranac started his career in motor-racing with a home-built 500 and some years later he designed the world-championship-winning Brabham racing cars.

Postscript

The history of 500cc racing would not be complete without a brief description of the 'modern' 500cc cars built on the same principles as the original cars back in 1946.

Credit for the first modern car should be given to Peter Voigt, who built the Voigt-Renwick Special in the early 1970s using up-to-date construction methods, materials and technology. The car was powered by a 500cc Konig marine engine. At first, Voigt's car competed against the elderly (by then they were twenty years old) historic 500s and, not surprisingly, was able to outperform them in no uncertain manner. It was not long before other aspiring racing drivers latched on to this idea and began to produce cars based on similar principles to Voigt's. The modern movement received a considerable boost when John Corbyn from Wellingborough, Northamptonshire, constructed a prototype car that he called the Jedi. So successful was this car that the interest it generated prompted Corbyn to go into production and by 2003 well over one hundred Jedis had been built, a number having been ordered by overseas customers. Other manufacturers such as OMS have produced similar cars, which resemble scaled-down Grand Prix cars with performance and road-holding to match. Modern 500s such as these are powered by multi-cylinder Japanese motorcycle engines such as Suzuki, Honda and Kawasaki. Modern 500cc racing is proving very popular and such cars are eminently suitable for use as circuit racers as well as continuing their original hill-climb and sprint activities.

A modern 500 – a Jedi, produced by John Corbyn's factory at Wellingborough, negotiates Ettore's Bend at the Prescott hill-climb.

Appendix: 500cc production racing cars

As soon as the popularity and success of 500cc racing had become established, various manufacturers offered series-production cars to the public. Overall, the Cooper emerged as the most successful and numerous of the factory-built cars, but the brand was bravely challenged by other makes who also enjoyed their fair share of success. Listed below are the main contenders.

ARNOTT

Better known as a manufacturer of superchargers, the Arnott factory at Harlesden produced a limited number of 500cc racing cars, the first Arnott making its appearance at Brands Hatch in June 1951, driven by George Thornton, its designer. Latterly the Arnott factory ran a racing team that achieved some modest successes. (Number produced – 9)

COOPER

Building production cars was not in father and son Charles and John Cooper's minds when they built their first 500cc-engined car, which made its debut at the Prescott hill-climb in 1946. However, after they had built a second 'prototype' for John's friend Eric Brandon and the duo began chalking up numerous successes, enquiries soon began to reach the Cooper garage in Surbiton. Initially, to cope with this demand, a small batch of twelve cars was produced. One of the first customers to be allocated one of these was Stirling Moss. This first batch of cars

A production machine, although not seen until 1951, was the Arnott, whose manufacturer was better known for its superchargers. The Arnott team, managed by Miss Daphne Arnott, put up some creditable performances. Gerald Smith is seen in October 1952 at a club meeting at Silverstone, but he was unplaced.

Left: *A well-known name in Formula Three racing, C. A. N. (Austen) May started racing in 1949 at Goodwood, having purchased Stirling Moss's first Cooper. May was a wholehearted 500cc enthusiast and was the author of 'Formula Three', published in 1951. He raced and hill-climbed his Coopers for many years and on one famous occasion made the fastest time of the day at the Prescott hill-climb. He is seen here at Silverstone in September 1954.*

Right: *A young Stirling Moss poses for the camera by his Cooper Mark 3 to display his winner's spoils.*

laid the foundation for a series of major successes and orders flowed in. The Cooper factory went from strength to strength and, before long, more ambitious projects were being undertaken, including the manufacture of the Cooper-Bristol Formula Two racing car and sports cars. The Cooper factory carried the rear-engined principle into Formula One and its world championship victories with Jack Brabham as works driver are legendary. (Number produced – approximately 320)

EMERYSON

A departure from the norm, the Emeryson had a front-mounted engine and front-wheel drive. This gave the car a fine aesthetic appearance with classic lines – an excellent example of a

The beautiful lines of one of the few front-engined 500cc cars can be seen in this picture of Peter Jopp at the wheel of his Emeryson at Castle Combe in April 1954. Emeryson produced a total of eight cars and well-known drivers included Paul Emery himself, Harold Daniell and Bob Cowell.

A determined-looking Cyril Hale in a front-engined Emeryson at Silverstone in 1954. Note the megaphone exhaust pipe.

'scaled-down' full-blown racing car. The performance and handling of the front-engined car confounded all the critics and the Emeryson achieved a number of well-earned successes. (Number produced – 8)

IOTA

This was an interesting concept that was promoted by a Bristol-based organisation that included some of the originators of the 500cc formula itself, notably Dick Caesar. The idea was to produce a basic chassis that would then be made available to would-be builders of 500s. This gave an individual the basis of his car and he was then free to build it to his own requirements in terms of suspension, steering, braking and motive power. A number of well-known 'specials' started life as an Iota chassis, including the Wasp, Aikens and Milli-Union. Subsequently Iota produced a small number of complete cars, with the buyer fitting his own choice of engine. (Number produced – 12 chassis and 9 complete cars)

JBS

The brainchild of Wembley speedway star Alf Bottoms, operating in Feltham, Middlesex, the JBS was a serious challenger to the Cooper. The debut of the car happened to fall on the occasion of another debut – that of the opening of Brands Hatch as a car-racing circuit. The potential of the car was immediately apparent and a number of important successes followed. However, the JBS project came to an end with the untimely death of Alf Bottoms in an accident during practice for the Luxembourg Grand Prix in 1951, although the cars continued to be

Clive Lones in one of the production Iotas at Castle Combe on 12th May 1951. There was only one 500cc race at that meeting, which Lones won. This must have been the only occasion on which Iotas took the first three places.

A JBS, which threatened to rival the Coopers for performance before the untimely death of constructor Alf Bottoms, as portrayed in a contemporary children's publication.

campaigned for some years, a notable driver being Don Parker, one of the most successful of all 500 drivers. (Number produced – approximately 18)

JP

From a small factory at Bellshill in Glasgow the only Scottish 500 was produced in small numbers. Most were sold to Scottish drivers by the manufacturer, Joe Potts, and took part mainly in northern and Scottish events. A number of local successes were achieved but incursions to the south produced little in the way of silverware. (Number produced – approximately 15)

A Scottish 500 – a small factory at Bellshill in Glasgow was the only one in Scotland to produce Formula Three cars. Records do not indicate the total number of cars produced by Joe Potts but it is considered to be around fifteen. One of the first – and notable – drivers of a JP was none other than Ron Flockhart, who went on to drive for BRM. Here G. H. Brown laps at Silverstone in August 1953.

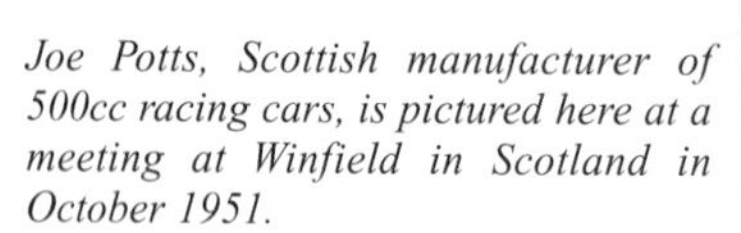

Joe Potts, Scottish manufacturer of 500cc racing cars, is pictured here at a meeting at Winfield in Scotland in October 1951.

Ken Gregory, secretary of the Half Litre Car Club, receiving a push-start in the swing-axle Kieft – probably 1951. The designer, Ray Martin, helps to push. Smoking was a lot more popular then!

KIEFT

The leading challenge to Cooper dominance, Cyril Kieft's company at Bridgend, latterly at Wolverhampton, produced cars to a revolutionary design incorporating the innovative features of rubber suspension and rear swing-axle and achieved some outstanding successes. With Stirling Moss as works driver and Don Parker as a privateer, the Coopers were vanquished on many occasions, Parker being National 500cc Champion on three occasions, driving a Kieft. The factory also succeeded in breaking international speed records at the Montlhery track in France. (Number produced – approximately 35)

MACKSON

Mackson Cars of Cobham, Surrey, although producing cars of identical build, did not actually offer them for sale to the public but formed a team of three cars, which made their first

Another manufacturer that went into production was Mackson, who ran a short-lived team of cars in 1952, with works drivers Gill, Braid and Shillito. Designed by Gordon Bedson, the cars did not make any noticeable impact on Formula Three events although a number continued to be raced in private hands.

Mike Trackman's Martin at Silverstone in 1960. Ray Martin was also prominent in the design of the Kieft and, not surprisingly, there is a marked similarity between the two makes.

appearance at Castle Combe in April 1952. They were placed in the heats and one of the cars, driven by Arthur Gill, finished fifth in the final. However, the Mackson team was short-lived and although Gill continued to campaign one of the cars as a privateer, the marque faded from the scene. As designed by Gordon Bedson, an aeronautical engineer, the driver sat very close to the front of the car over the centre of gravity. (Number produced – 3)

MARTIN

Ray Martin, one of the designers of the successful Kieft, went on to build his own production run of cars from premises in Merton, London. Unsurprisingly, the Martin was very similar to the Kieft, with tubular chassis and rear swing-axle. (Number produced – approximately 10)

MARWYN

Contrary to popular belief, Cooper was not the first series-production make to be marketed to the public. That distinction fell to Marwyn, a small organisation based in Bournemouth and subsequently at Wareham in Dorset. The prototype Marwyn made its first appearance at Brighton speed trials in September 1947 in the hands of Sir Francis Samuelson and came third

The prototype Triumph-engined Marwyn of Sir Francis Samuelson, seen here at Prescott in 1948.

A motor-racing circuit that ran outside the Grand National course at Aintree, Liverpool, was the venue for several 500 races. An Erskine-Staride driven by B. A. Manning receives a push-start at the July 1955 meeting. Note the extremely far forward location of the driving seat.

in its class. The prototype was fitted with a Triumph twin engine. A number of Marwyns were sold but were often extensively modified to the owners' requirements. (Number produced – 6)

STARIDE

Mike Erskine of Southampton designed and built a machine that was most distinctive in appearance, with a long body and a very forward driving position. The Staride and Mackson had these features in common. The Staride made its first appearance at Goodwood in April 1952 in the hands of experienced driver John Habin. He succeeded in bringing the car into fourth place in the Earl of March Trophy out of a field of seventeen starters. (Number produced – approximately 10)

Arguably the ultimate swing-axle design: speedway constructor Mike Erskine's Staride. This example is that of Dennis Taylor at Castle Combe in April 1954.

Further reading

The definitive works on the subject of 500cc motor-racing are *Formula Three* by C. A. N. May (Foulis, 1951) and *500cc Racing* by Gregor Grant (Foulis, 1950), but these have long since been out of print. Neither of these books is easy to find nowadays, but it is well worth acquiring a copy should the opportunity arise. Articles and references to 500cc racing appear from time to time in various motor-sport publications, and there is a club magazine, *The 500*, produced by the 500 Owners' Association (details can be found on the club's website: www.500race.org).

Places to visit

Several of the well-known motor museums in the United Kingdom often have an example of a 500 among their exhibits, but there is no museum dedicated specifically to this car. An example can usually be seen at the National Motor Museum (John Montagu Building, Beaulieu, Brockenhurst, Hampshire SO42 7ZN; telephone: 01590 612345; website: www.beaulieu.co.uk) and 500cc racing cars can often be seen in action at hill-climb venues such as Shelsley Walsh, Worcestershire, and Prescott, Gloucestershire. In addition, the 500 Owners' Association itself organises a hill-climb at Wiscombe Park, Devon, in May of each year. An up-to-date listing of all road-transport museums in the United Kingdom can be found on www.motormuseums.com

Organisations

The 500 Owners' Association. Website: www.500race.org
The Cooper Car Club Ltd. Website: www.coopercars.org

Reg Bicknell in his distinctive semi-aerodynamic Revis in the Castle Combe paddock on 3rd April 1954, the first meeting of the season. He had a good day on that date, winning both the first heat and the final from an entry of some thirty cars.